Sea Otters and Seaweed

How do we know what we know about the life histories of animals? Focusing on the sea otter, a mammal that lives along the edge of the North Pacific, the author describes the long-term, patient observations made by biologists. The resulting data create a picture of the life history of the sea otter, an animal that is, in many ways, unique. This picture forms part of a still bigger picture — the ecological role of the sea otter in the giant kelp forests of the North Pacific. *Sea Otters and Seaweed* opens a window on the lives of sea otters and the work of scientists that study them.

Sea otter in a bed of giant kelp hammers a clam on a
rock to open it.

SEA OTTERS and SEAWEED

By Patricia Lauber

GARRARD PUBLISHING COMPANY
CHAMPAIGN, ILLINOIS

Photo Credits

William F. Bryan: pp. 10, 53
Ron Church/Photos Unlimited, San Diego: pp. 9,
    13, 14, 15, 16, 20, 26, 46, 63
Jeff Foott from Bruce Coleman: pp. 2, 6, 31, 34,
    49, 54
Karl W. Kenyon from National Audubon Society
    Collection/Photo Researchers: pp. 19, 28,
    41 (both), 43
Patricia Lauber: p. 23
Tom McHugh from Photo Researchers: p. 33
Cover Photograph by Ron Church/Photos Unlimited,
    San Diego.

Library of Congress Cataloging in Publication Data

Lauber, Patricia.
    Sea otters and seaweed.

    1. Sea-otters.   2. Kelp.   I. Title.
QL737.C25L38   599'.74447   76-17796
ISBN 0-8116-6106-7

Contents

While resting, a sea otter may anchor itself with kelp. The kelp is fastened to the ocean bottom and keeps the otter from drifting. Sea otters are members of the weasel family. Closest relatives are river otters.

1. A Home in the Sea

The sun has just risen. Its rays light up a rocky coast. This is Point Lobos, in central California. The water shines. Offshore a big bed of seaweed glitters as it bobs in the waves. It is the kind of seaweed known as giant kelp.

Several sea otters are floating near the kelp. They are on their backs, and they look a little like logs. Their bodies are long and smooth and brown.

One otter has been resting. He floats with a piece of kelp over his chest. The kelp is anchored to the ocean bottom. It keeps the otter from drifting out to sea. Now the otter slides out from under the kelp and swims away.

To move fast, a sea otter rolls onto its belly. It swims by kicking with its webbed hind feet. But usually an otter swims the way this one is doing. He is floating on his back and using his tail as an oar. For a little extra speed, he kicks with a hind foot.

Several otters are grooming themselves.

Another rolls over and dives with strong kicks of his hind feet. In less

An otter dines off his chest. A fully grown male is four to five feet long and weighs 80 to 100 pounds. Females are smaller.

than a minute he is back with a crab. He uses his chest as a table to hold his food and begins to eat. He stops only to roll in the water and rinse food out of his fur.

Some distance away, there is a mother otter. She floats with her small pup on her chest. The pup is nursing. The mother is grooming its fur.

Mother otter grooming pup on her chest

Furred animals that nurse their young on milk are called mammals. Most mammals are land animals. But sea otters are mammals that live in the ocean. Their home is in the North Pacific.

All mammals are warm-blooded. To live in the sea, they must have special ways of keeping warm. Water carries away heat much more quickly than air does.

Most sea mammals have a thick layer of fat under their skins. It seals in much of their body heat.

Sea otters do not have this fat. An otter depends instead on its fur. The fur does two jobs. It keeps water out, and it holds heat in.

The fur is so thick that it is waterproof. Only the tips of the hairs get wet. The underfur stays dry, and water does not touch the otter's skin.

Trapped among the hairs are millions of tiny air pockets. Together, the fur and air pockets hold heat in.

Only clean fur can keep an otter dry and warm. Dirty fur becomes matted. Clumps of hairs pull apart, letting water in and heat out. To stay alive, sea otters must groom themselves often. They scrub and comb and fluff their fur with their feet. Sea otters also need to get air into their coats. They do this by turning, rolling, and somersaulting in the water. Very often they blow air into their fur.

An otter may shake itself and roll to rinse its coat. The bubbles put air in the fur.

The bottoms of an otter's paws are bare. These hairless surfaces get cold. An otter warms its paws by rubbing them together, just as people warm their hands.

To make heat, a mammal's body needs fuel. The fuel that is used is food. Sea otters have big appetites. An

80-pound otter eats about 16 pounds of food a day. This is a fifth of its own weight. It may, however, eat less one day and more the next.

Otters eat about 40 kinds of sea animals. They eat octopuses. They eat squid, which are related to octopuses. Some eat fishes from the ocean bottom. But mostly they eat sea

A sea otter may swim by kicking with one or both of its webbed hind feet.

After a successful dive, this otter eats a squid while holding the next course, an abalone, with one of its hind paws.

animals with outer shells, such as clams, mussels, crabs, and snails. They eat sea urchins, which have round shells covered with spines. California otters eat abalone, which are like giant snails.

Coming up from a dive with a large sea urchin

To get its food, an otter first takes many deep breaths. It swims toward the bottom and gathers its food. Then it surfaces to breathe. The deepest that otters can dive is about 160 feet. That is why sea otters are usually found close to shore.

An otter gathers food with its small front paws. It also uses these paws to carry large items of food. To carry small food items, an otter does something else. It makes a pocket for them with the skin of its chest. A sea otter's skin is loose on its body. If an otter pushes one front paw into the fur of its chest, the skin forms a deep fold, or pocket. The otter can cram 16 or 20 snails into this pocket.

Surfacing, the otter rolls onto its back. It plucks out a snail, and it cracks the shell with its paws or teeth. It scoops out the soft insides and eats them. Then it gets rid of the shell and rinses its fur. It plucks out another snail.

A sea otter can also open a small clam with its paws or teeth. But some California clams have big, hard shells. If an otter finds a big clam, it brings up a rock too. It uses the rock as a tool to get at the clam meat. The otter may place a flat rock on its chest and hammer the clam against the rock. Or it may place the clam on its chest and use the rock as a hammer.

To open his clams the otter is hammering them against a rock on his chest.

An abalone has one foot, which it can use as a suction cup. A sea otter gathers an abalone by pounding it with a rock.

Sea otters also use rocks to gather abalone. A big abalone has a single shell about seven to nine inches long. The shell is hard. And the abalone anchors itself firmly to the rocky ocean bottom.

A human fisherman must use a tire iron to pry an abalone loose. A sea otter picks up a large rock and hammers the abalone. Sometimes the shell breaks. Sometimes the whole abalone is knocked loose. The sea otter carries its find to the surface and dines off its chest.

By eating well and keeping clean, sea otters can live in the ocean. Like other sea otters, the ones at Point Lobos are very much at home among the waves and kelp.

They are also at ease because they do not know that they are being watched. But on shore there are several people with field glasses and small telescopes. They have been at Point Lobos since dawn, for early morning is a time when sea otters are very active.

2. The Otter Watchers

The chief watcher is a scientist named Judson Vandevere. Everyone calls him Jud. Jud's field of science is sea life, or marine biology. His special interest is sea otters.

The other watchers are students, aged 16 to 50. They want to learn about sea otters. They also want to learn how scientists like Jud work.

These scientists are men and women who study animals in the wild. They watch animals for years and keep careful records. Bit by bit, they put together a picture of how the animals live. That is how the rest of us learn about the animals.

The scientists are usually hidden. They want to know how animals behave naturally. Jud goes near sea otters only to take photos. The rest of the time he watches from the shore.

Jud searches for a particular otter from a high place at Point Lobos State Reserve.

Once he studied a mother and her pup from sunrise to sunset, rain or shine, every day for three weeks.

Now he and the students have been watching another mother and pup. They write down what the mother does and what the pup does. Is this pair different from other pairs Jud has studied? The records will show the answer.

The mother they have been studying has a tag on her left hind foot. That is how they can tell her from other mothers. The tag was put on by the California Department of Fish and Game. Tagging wild animals is a way of finding out where they go and what becomes of them.

Yesterday the students couldn't find the mother. They searched up and down the coast without seeing her. Today they are hoping she will turn up at Point Lobos. She has often been here before. They hope the mother and pup by the bed of kelp are the pair they want.

Several hours go by. The sun creeps higher in the sky. Just before ten o'clock the mother otter swims by kicking with her hind feet. The left one comes into view. There is the tag. They have found her!

Jud is seated behind a small telescope. A student sits beside him with a record form. She fills in the blanks for date, place, and weather.

She starts recording what Jud reports. When he started watching, the mother was grooming the pup's neck. The pup was playfully biting its mother's face.

Now the pup has started to nurse. Jud gives the time, which is 10:01.

Sometimes it takes practice to be able to tell the otters from the seaweed.

The pup has its head near the mother's back legs. That is where her nipples are. While the pup feeds, the mother grooms its hind end.

The nursing period lasts four minutes. The pup tries to go on nursing, but the mother turns it away. She rolls over and begins to groom herself. The pup floats in the water beside her and watches.

Between reports there are long waits. During them Jud talks about mother otters and their pups.

A female sea otter, he says, is three or four years old before she can mate and have young. So far, no one has seen the birth of a California sea otter. But Jud once saw a mother

Mother sea otter holding her pup

with a newly born pup. They were in the water, and he thinks births take place there.

A pup is born with a long coat of thick, woolly, yellow-brown fur. It is a big baby, weighing about five pounds. Its eyes are open, and it can float by itself. But it must learn to swim and dive and groom itself. It does this by watching its mother and trying to do what she does. At first it cannot. Its muscles are too weak, and they do not work well together. A sea otter pup is about a year old before it can really take care of itself.

Jud reports: "The pup's body is in the water to the right of its mother. But its head, arms, and chest are on the mother's chest. The pup is playfully biting her."

The pup Jud is watching is about one month old. Its coat is still very woolly. It eats a little bit of solid food that its mother places on its chest. But most of its food is milk.

The pup has learned to swim. It can swim to its mother. It chases seagulls. It sometimes watches what its mother does underwater, but it is not yet able to dive. It can get its head underwater. But the rest of its body floats like a cork while its hind feet paddle in the air.

Jud reports: "The mother is rubbing her webbed hind feet together."

Even when a pup can dive, it still does not find much of its own food.

At first it brings up stones or empty
shells with no value as food. During
its first year a pup gets most of its
solid food from its mother.

An otter with a sea urchin attracts a gull which hopes
for scraps.

Jud reports: "The mother is grooming the pup. The pup is moving toward a nipple and being pulled back. Now it's being allowed to nurse while the mother grooms its lower back.... The pup has stopped nursing. Asleep? No, it has changed nipples.... The mother just pulled the pup away from her nipples. The nursing period was 7 minutes 36 seconds. The pup changed nipples twice."

This mother otter has some unusual feeding habits, Jud says. Her feeding periods are very long. They may go on for 4½ hours. Also, her dives last

longer than those of other otters—
from 2 minutes to 2 minutes and 45
seconds. Jud has timed only one
other sea otter that equaled this
record. This was a male otter with
a big chest. A third interesting
thing is that this mother eats
mostly octopuses. She seems to be

Even a big pup stays with and depends on its mother.

A mother sea otter tends her sleeping week-old pup in a bed of kelp.

an expert at knowing where to find them under rocks.

When the mother dives for food, she leaves the pup at the surface.

Jud reports: "Now the pup is examining its mother's fur. It may be sleepy—it's nodding. The mother has turned the pup so that its face is under hers. . . . Now the pup is lying on its back, rolled into a ball on its mother's chest."

Sometimes a mother and her pup are separated during her feeding dives. The tide may sweep the pup away. Or the mother's dives may carry her away from the pup. As soon as the pup misses its mother, it calls with a sharp, shrill cry. The mother hears the cry and looks for the pup. If she does not see it, she calls back. The pup cries again. The search goes on as the mother calls and the pup answers. When she finally finds the pup, it stretches out its paws. The mother hugs it to her chest.

But sometimes a mother cannot find her pup. The pup may starve to death. It may be taken by a bald eagle, a man, or a shark. It may be dashed against a rock. The mother may also die. She is likely to have spent her time searching for the pup instead of feeding.

Jud reports: "This is the closest the mother has been to other sea otters since we've been watching her. . . . Oh, all three of the other otters have pups —that's why she let them come so close. But the other pups are bigger than hers."

The day wears on. The record sheet fills up. It will be added to Jud's other records. Together these records tell much about California's sea otters. They let Jud draw a word picture of the otters' way of life. He can also compare this with the picture drawn by other scientists of sea otters that live to the north.

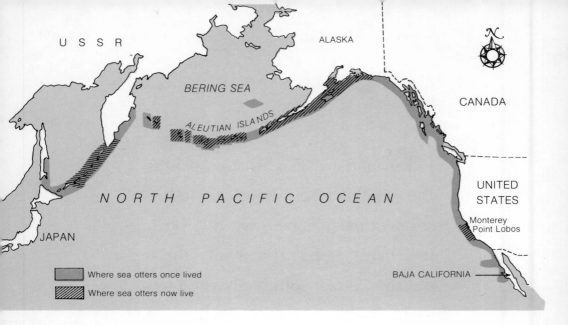

Map labels: U S S R • ALASKA • BERING SEA • ALEUTIAN ISLANDS • CANADA • NORTH PACIFIC OCEAN • UNITED STATES • Monterey • Point Lobos • JAPAN • BAJA CALIFORNIA

Where sea otters once lived
Where sea otters now live

3. Otters, North and South

Once there were many sea otters along the edge of the North Pacific Ocean. Then, in the 1700s, explorers discovered the otters and their thick, soft fur. Word spread of the riches to be had in furs. Hunters poured in and began killing every sea otter they could find. By 1900 the otters had been nearly wiped out.

In 1911 several nations signed a treaty that stopped the killing. By then there were few otters left. So far as anyone knew, all of these were in the far north.

Left alone, the northern sea otters made a comeback. Today there are between 60,000 and 100,000 of them.

In 1938 a colony of sea otters was discovered off the coast of central California. A few southern sea otters must have escaped the hunters. They multiplied and became a colony. Today about 1,400 otters live along a 160-mile stretch of California coast.

The northern and southern otters live in very different places. The north is much colder and has fiercer storms. It also has few people. How does the northern otter's way of life compare with the southern otter's? There are many differences.

Northern otters often leave the water, or haul out. They may haul out onto beaches or rocks to rest. They may haul out to escape the winds and waves of winter storms.

Hauling out is rare among southern sea otters. Some scientists think these otters never hauled out much because of danger from bears, mountain lions, coyotes, and man.

Perhaps because they are used to hauling out, northern otters can move easily and quickly on land. They can travel on all four legs.

Northern sea otters are much more at home on land than southern sea otters are. They often haul out, and they move easily on land.

A southern sea otter is awkward on land. It may walk on its front legs and drag its hindquarters. Or it may move by swinging its body up and forward.

Neither northern nor southern sea otters form family groups. A male and a female come together for a few days to mate. Then they separate.

In the north a male and a female come together near a rock chosen by the female. They spend about three days near this rock, mating, hauling out, and feeding together. For the first two days both otters find their own food. When the female dives, the male follows her closely. He surfaces a second after she does, and they feed,

A sea otter pup nurses with its head near the mother's hind legs. Northern otters often nurse their young when hauled out.

floating close together. But on the third day the male starts stealing food from the female. He quickly eats his own food and then takes food from his mate's chest and mouth.

Soon after, the female brings the mating period to an end. This is how it happens. Because the male is bigger, he needs more food. He sometimes gathers food when the female does not. She waits until his diving carries him away from the rock. Then the female suddenly leaves the rock and swims away underwater. When the male notices that she is gone, he searches for her. But he usually does not find her.

In the south mated pairs of otters also spend three days together. But they spend their time in or near a bed of kelp. When they rest, the male may lay his head on the female's chest, as if he were a pup.

Southern males steal food from their mates on all three of these days. Jud once watched a male that dove and surfaced with his mate. The male would bolt his own food and then take food from the mouth, paws, and chest pouch of the female. If she was using a rock to open snails, he took both her rock and her snails. Having lost her food, the female would dive again. The male always followed her, even when he had not yet finished his stolen food.

In the south the female also ends the mating period by slipping away.

There are many other differences between northern and southern sea otters. There are also many ways in which they are alike.

These studies of sea otters are of interest to another group of marine biologists. They are trying to draw a still bigger picture—the life of the kelp forests.

Fronds of giant kelp are kept afloat by air bladders the size of golf balls. Giant kelp is one of the longest plants in the world and may grow 200 feet in length.

4. The Otters and the Kelp

Sea otters can live where there are no beds of giant kelp. But they prefer to live near the big seaweed. Here they can anchor themselves while resting. Here also there is plenty of food. Many kinds of shellfish and other sea animals live in or near the giant kelp. They, too, find food and shelter in the kelp.

That is why a bed of giant kelp is sometimes called a forest of the sea.

Like any forest, it is a neighborhood of plants and animals that live together.

The giant kelp takes the place of tall trees in this forest. The kelp is anchored by its holdfasts. Bundles of ropelike stems grow from the holdfasts. The stems reach upward toward the surface, which may be 30 to 125 feet above. Long streamers grow from the stems. These are called fronds. They spread out and float at the surface.

The big fronds shade the depths below. But there are openings where light breaks through. Where there is light, other plants can grow. There are small seaweeds. There are still smaller relatives that grow like moss on rocks.

Sea snail on giant kelp

Animals are all around. Small crabs and mussels and worms hide among the holdfasts. Snails are everywhere, crawling through the holdfasts, along the stems, and over the fronds. Fishes of all sizes dart and float among the seaweeds.

As in any forest, the animals depend on the plants for food. The seaweeds and their relatives are all green plants. They may not appear green. But the green is there, hidden by some other coloring matter. Like all green plants, these make their own food. And so they are the starting point of food chains in the forest.

Abalone, for example, feed on bits of seaweed that have broken off. Snails graze seaweed by scraping it with their file-like tongues. Other animals prey on the plant eaters. An abalone may be eaten by an octopus. The octopus, in turn, may be eaten by a sea otter. So may both abalone and snails.

In a healthy kelp forest, there are many kinds of plants and animals. Each kind acts as a check on others. And so the life of the kelp forest stays in balance.

For example, suppose the giant kelp were left alone. It might crowd out other forest plants by cutting off their light. But this does not happen. Grazing animals feed on the giant kelp, cutting it back, trimming it, and making holes in its fronds. They do the same thing to smaller seaweeds, giving light and space to the small plants that grow on rocks.

Suppose the grazing animals were left alone. They might harm the plant life by eating too much. But this does not happen either. In a healthy kelp forest, there are many animals that prey on the plant eaters.

Marine biologists have been studying the kelp forests of the North Pacific. Some think that sea otters are the key to healthy forests. They reason this way.

Sea otters eat many kinds of sea animals. They feed both on plant eaters and on animal eaters. They need a lot of food to fuel their bodies. And they need food that they can find easily. Otherwise they must burn up fuel in searching for food. So otters usually eat whatever is most plentiful.

Suppose one kind of sea animal is

multiplying rapidly. In time it might crowd out other forms of life. But it cannot if otters are around. The sea animal becomes a plentiful source of food—and the otters feast on it. Their feasting brings the population under control.

An otter is scooping out the soft insides of a purple sea urchin, a favorite food.

Sea urchins feeding on kelp holdfasts

Sea otters are very important in controlling sea urchins.

Usually, adult sea urchins live in big cracks in ocean rocks. They feed on bits of plant matter that drift by. But sometimes there is not enough drifting food. Then the urchins leave their cracks. They move in on the giant kelp itself. Their sharp teeth can chew up fronds near the base of the plants. They can cut through stems. If they do, whole plants float away. With them go food and shelter for many forest animals.

In a forest with sea otters, this is not likely to happen. Once the urchins leave their cracks, they become easy to catch. And they are one of the foods that sea otters like best.

What happens to kelp forests in oceans that have no sea otters? Some marine biologists think that other animals are the key to health in these forests. Along the Atlantic coast of North America, the key animal may be the lobster. It also preys heavily on sea urchins.

The health of kelp forests is important to people. In some places, such as Japan, seaweed is a food that many people eat. In other places, such as southern California, there are companies that harvest kelp. Giant kelp holds a useful chemical called algin. It is used in the making of many products. Among these are ice cream, writing paper, car polish, paint, cement, salad dressing, and toothpaste.

A kelp forest is a nursery for young fishes and shellfishes. It also helps animals that do not live in it. It produces food for them.

Giant kelp is the fastest growing plant on earth. It grows at a rate of about 18 inches a day. Twice a year parts of the giant kelp die and are shed. Huge amounts of plant matter are carried away from the forest. They become food in other neighbor-hoods of the sea and shore.

Some of our seafood lived in a kelp forest when it was young. Much more of it was part of a food chain that began in a kelp forest. The forests play a part in all our lives.

So do sea otters, if they are the key to healthy kelp forests in the North Pacific. Yet the southern sea otters face dangers that threaten their future.

5. Today and Tomorrow

The population of California sea otters is small. All the otters live along one piece of coast. And this is a place where there are many human activities. Those are three reasons why the otters are in danger today.

The threat of an oil spill always hangs over them. One big spill where they live could kill most of them.

Factory wastes are another problem. Many of these hold chemicals and metals that are harmful to animal life. Yet these wastes reach the sea in sewage.

Heated water is a danger. Power plants get rid of heated water by letting it flow into the sea. This water is not good for either sea otters or giant kelp.

Sea otters are killed by boats. They are sometimes shot by people who don't like them. Mostly these are fishermen who are after the kinds of shellfish that both people and otters like to eat.

Because of these dangers, some people fear that the southern sea otters may be wiped out. To safeguard the otters, they would like to start a

Sea otter with abalone, which is also prized by human fishermen.

second colony. They would like to move some otters to a place with clean water, no oil tankers, and no human fishermen.

The trouble is that moving sea otters costs a lot of money. It is hard to keep them alive. Even if an otter is kept clean, wet, and cool, it may become so upset that it dies.

Will a new colony take hold and do well? No one yet knows. A few years ago 180 northern sea otters were moved from a place chosen for an atom bomb test. Sixty were taken to British Columbia, 60 to Washington, and 60 to Oregon. But it is too early to tell how these colonies are doing.

Still other things could be done to

help the southern sea otters. Perhaps they will, for otters give pleasure to thousands of visitors.

Also, many people are starting to understand something important. This is that all living things are bound together in webs of life. Human beings are part of those webs. For our own sakes, we must learn to share the earth with other forms of life. And we must learn now. Tomorrow may be too late.

At Point Lobos, Jud and his students are packing up. The sun is setting. Its slanting rays make the water shine. The kelp bed bobs and glitters.

One sea otter stands up. Treading

water with its webbed hind feet, it takes a look around. The students look back and smile at the sight.

All the otters are active now. Diving and feeding, they are using the kelp forest. But they are also part of it, and they give something back.

Index